Emerging Technologies in Clean Energy 2024

Topical Conference at the 2024 AIChE Spring Meeting and 20th Global Congress on Process Safety

New Orleans, Louisiana, USA
24-28 March 2024

ISBN: 979-8-3313-1711-9

Printed from e-media with permission by:

Curran Associates, Inc.
57 Morehouse Lane
Red Hook, NY 12571

Some format issues inherent in the e-media version may also appear in this print version.

Printed with permission by Curran Associates, Inc. (2025)

For permission requests, please contact AIChE
at the address below.

AIChE
120 Wall Street, FL 23
New York, NY 10005-4020

Phone: (800) 242-4363
Fax: (203) 775-5177

www.aiche.org

Additional copies of this publication are available from:

Curran Associates, Inc.
57 Morehouse Lane
Red Hook, NY 12571 USA
Phone: 845-758-0400
Fax: 845-758-2633
Email: curran@proceedings.com
Web: www.proceedings.com

TABLE OF CONTENTS

ALTERNATIVE FUELS INCLUDING BIOFUELS, HYDROGEN, RENEWABLE HYDROGEN, AND SYNGAS IV

ALTERNATIVE FUELS INCLUDING BIOFUELS, HYDROGEN, RENEWABLE HYDROGEN, AND SYNGAS III

76c High Pressure Fluidized Bed Reactors

Eloy Flores III, Veshal Venkat

At SwRI, we engage in the process development and improvement for a variety of sustainable fuels and chemicals from lab to pilot scales. The applied research, development, testing and evaluation methods undertaken help bridge the gap between fundamental science and commercialization. The technologies developed include but are not limited to hydrogen production, CCUS, renewable fuels, chemicals and materials production and conventional thermos, photo and electro-chemical conversions. One such technology is the High Pressure Circulating Fluidized bed (HPCFB) reactors. HPCFB reactors have recently showcased itself as a promising technology for sustainable fuel and chemical conversions. These reactors have several advantages over conventional reactors, including higher rates of heat and mass transfer, improved mixing of reactants, and the ability to recycle catalysts with high recirculating solid flow rates for heat recovery. HPCFBs are particularly suitable for high-temperature, high-pressure processes such as gasification, pyrolysis, and hydrotreating. At SwRI, we have developed HPCFBs and the supporting infrastructure for practical applications, ranging from small-scale laboratory experiments to larger pilot-scale CFBs. These reactors can efficiently convert both renewable and petrochemical feedstocks into valuable end products. The reactor system has also shown promise for the one-step cracking, deoxygenation, and hydro-isomerization of biogenic feedstocks into specific fuels and chemicals.

97a Invited Talk: Overview of Doe/Fecm Clean Hydrogen Programs

Jai-woh Kim

Clean hydrogen can be produced from various fossil fuels and solid waste feedstocks with CCS or low-carbon energy sources using electrolyzers and can be utilized without emissions greenhouse gases. Several areas with the hardest-to-abate emissions like steel, cement, refinery and chemicals can be decarbonized with hydrogen.

The U.S. Department of Energy's Office of Fossil Energy and Carbon Management (FECM) is focusing on development of carbon-neutral hydrogen production from natural gas and advanced innovative designs and technologies for converting diverse types of solid waste feedstocks (MSW, non-recyclable plastics, waste coal) with biomass into clean synthesis gas to enable the low-cost production of clean hydrogen with the integration of pre-combustion carbon capture (CCS) to achieve net zero carbon emissions.

This presentation will provide a broad overview and progress that the DOE's FECM Gasification Systems and Reversible Solid Oxide Fuel Cell (RSOFC) Programs has initiated on recent work supporting from various waste feedstocks with biomass and cost reduction of each component of the overall gasification systems and from materials development and small scale demonstration of RSOFC systems along with a detailed discussion of future low-cost technology advances for clean hydrogen production.

97b Hydrogen Production from Dimethyl Ether Steam Reforming (DME-SR)

Xiaokun Yang

Dimethyl ether (DME) is, by many performance and cost metrics, the best multipurpose renewable fuel and hydrogen carrier. DME boasts a hydrogen gravimetric capacity of 13 wt.% and a volumetric capacity of 87 kg H_2/m^3 and is ideally suited to be used as a hydrogen carrier in support of the hydrogen economy. We developed a novel structural catalyst as a micro-reactor to improve hydrogen selectivity (95%) and yield from dimethyl ether-steam reforming under 300 °C. The process had been demonstrated in lab-scale and supported by TEA/LCA analysis. Long-term strategies to produce DME from sequestered CO_2 offer a promising approach to develop DME as a carbon neutral hydrogen carriers.

173a Identification of Non-Aqueous Solvents for Pairing with Amine-Type CO_2 Capture Systems

Adhish Chandra Saketh Madugula, Aishwarya Roy, Tracy Benson

The Industrial Revolution of the 19th century unlocked the formidable potential of fossil fuels, establishing their enduring dominance as the primary energy source up to the present day. Since 1950, the consumption of fossil energy, harnessed from the combustion of oil, gas, and coal, has grown eightfold, constituting nearly 84% of the global energy mix in 2019. Although these fossil fuel-based energy feed stocks have consistently driven economic prosperity across nations by energizing diverse industries and elevating living standards for an expanding population, a noteworthy drawback lies in the substantial increase in the atmospheric carbon dioxide (CO_2) emissions resulting from the combustion of these fuels. Elevated atmospheric CO_2 levels pose a dual threat: a) contributing to the greenhouse effect with potential long-term climatic repercussions and b) posing direct health risks due to increased air pollution.

In view of these challenges, carbon capture and storage (CCS) has positioned itself as the most sought out technology to abate CO_2 emission from large point sources. Particularly for post combustion processes, amine absorption is the preferred method for capturing waste CO_2. Current amine absorption technology typically uses 30 wt% monoethanolamine aqueous solution. This solvent is considered as the most mature amine scrubbing solvent available for CO_2 capture due to its advanced technological readiness, ease of operation, impressive CO_2 absorption rates, and recyclability. However, CO_2 capture via aqueous MEA imposes a substantial energy burden during solvent regeneration, thereby driving up the overall cost to capture CO_2.

The desorption of captured CO_2 during the solvent regeneration phase necessities heating the CO_2-rich aqueous solvent to near boiling temperatures (80-120 °C). Given that water accounts for ~ 70 wt% of the aqueous amine solvent, the majority of the heat being supplied by the reboiler of the regeneration column is associated with the heating and vaporization enthalpy of water. Therefore, adopting of non-aqueous solvents with lower heating and vaporization enthalpies would result in reduced heat requirements in the regeneration column, thereby making the CO_2 capture process a potentially compelling, low-energy, and cost-effective technology.

This work endeavors the development of a multi-parametric framework using Aspen simulations and optimization tools, to screen and identify potential non-aqueous absorbents that are compatible with amine-type solvent while also being suitable for post-combustion flue gas conditions. Some of the potential non aqueous solvents discussed in this work include ethanol, triethylene glycol, and polyethylene glycol. These solvents were screened based on parameters such as normal boiling point, viscosity, CO_2 loading, relative volatility, enthalpy of vaporization, enthalpy of CO_2 absorption, MEA and H_2O compatibility, regeneration efficiency and thermal stability.

173b Strategies for Decarbonizing the Chemical Process Industry

Jose A Aguilar, Michael Hartmann

The growing demand for technologies to capture, utilize, and sequester CO_2, bolstered by government and public realization of meeting global energy demands and climate goals, is an opportunity for the chemical and process industries. Existing facilities around the world have the potential to capture more than 40 $MtCO_2$ every year with the advent of capture technologies in the 1970s for use in enhanced oil recovery. There has been a building momentum in the field arising from investment incentives and the civic value of meeting climate targets. The technologies that have been implemented the most involve chemical absorption or physical separation, which have worked successfully for decades now. At SwRI, we have integrated teams working on all facets of CCUS and related infrastructure developed for real-world CO_2 producers. Our expertise ranges over chemisorption, membrane and oxy-fuel separations, chemical and calcium looping, direct air capture, infrastructure and logistics, and supercritical CO_2 power cycles. SwRI has a notable share in the development of technologies for the approaching net-zero carbon economy by 2050 and a 100% clean electrical grid by 2035.

We also work in direct decarbonization of natural gas streams to hydrogen and value-added carbon-containing product fractions. Most of the ongoing work involving low-carbon fuels, sustainable aviation fuels, and hydrogen from feedstocks like biomass, waste plastics. and municipal waste comprise directly and indirectly SwRI CCUS advancements. We also have expertise in syngas production and upgradation, CO_2 conversion to fuels and chemicals, fermentation, and electrolysis. We strive to advance CCUS technologies through fundamental research, process development, and industrial application. Through numerous and diverse projects completed over decades SwRI promotes an achievable and sustainable energy and material economy. This presentation will cover the different approaches to decarbonization in the industry.

173c Process Modeling of Pre-Combustion CO_2 Capture Using Novel Ionic Liquids for Blue H_2 Production: 3E Evaluation

Sadah Mohammed, Fadwa Eljack, Kazi Khoda

With the imperative shift toward a low-carbon economy and the goal of achieving net-zero emissions by 2025, the global demand for hydrogen production is expected to rise. It is projected that the demand for hydrogen to reach almost $12 trillion by 2050 [1], creating opportunities for countries with abundant natural gas reserves to export environmentally friendly hydrogen. Hydrogen, a non-carbonaceous fuel, is an alternative to conventional fossil fuels like natural gas. It has low carbon emissions and relatively high power density, and it can also be blended with natural gas to meet the fuel demand of different energy systems [2]. More than 95% of the worldwide hydrogen demand is satisfied via reforming fossil fuel, with steam methane reforming (SMR) accounting for 50% of the total production (Oh et al., 2022). However, the SMR process, being energy intensive, leads to high CO_2 emissions (89.1 gCO_2/MJ) [3]. To overcome this challenge, blue hydrogen production, where the CO_2 is removed from hydrogen effluent gases in fossil fuels processes using CO_2 capture technology is a viable solution with lower carbon intensity (22.4 gCO_2/MJ) and offers a feasible route to sustain reliance on fossil fuels [5], [6]. In the context of pre-combustion, CO_2 removal from processes typically employs absorption methods [7]. The exploration of chemical absorption-based processes in hydrogen production systems aims to create versatile, energy-efficient, and cost-effective pre-combustion CO_2 capture systems [8]. Ionic liquids (ILs) have gained popularity due to their favorable CO_2 capture properties in pre-combustion applications, such as low vapor pressure [9]. The primary goal of this work is to enhance efficiency and reduce capital and operating costs associated with pre-combustion CO_2 capture for blue hydrogen production, utilizing novel ILs. These ILs are designed using a predictive deep learning model (DL) developed by our research team in previous work [10]. The 'PySCF' package in the 'Python' programming language is employed to optimize the ILs' geometry. Subsequently, the optimized geometry is input into COSMO-RS software to acquire essential data, including sigma profile (SGPRF1 to SGPRF5), activity coefficients, etc., to define the novel ILs in the simulation software 'Aspen Plus' following a COSMO-based/Aspen approach. Once the ILs are defined, a pre-combustion CO_2 capture process involving packed absorption and flash for solvent regeneration is developed in Aspen Plus. Finally, the study examines and compares the process system engineering aspects, encompassing energy, exergy, and economics (3E), of the developed system with established ILs. This research is pertinent to existing SMR facilities, ensuring high purity in produced hydrogen and mitigating CO_2 emissions. Furthermore, the 3E analysis is vital for enhancing the overall process viability of ILs for pre-combustion CO_2 capture and gaining insights into the application of novel ILs in large-scale plants.

154b Thermal Efficiency and Heat Recovery in the SNG Synthesis System

Alejandra Peralta, Yoshitsugi Kikkawa, David Hill

Because of the increase in global warming and the need to find new ways to minimize carbon footprint, the utilization of renewable energy has become essential in recent years.

In previous projects, we proposed an approach to utilize renewal hydrogen energy and Capture Carbon Dioxide to produce Synthetic Natural Gas (SNG) by classic Sabatier reaction. This SNG is produced by the methanation of CO2 from fossil fuel power stations and hydrogen produced by water electrolysis.

The SNG is transported by pipeline to LNG tankers to the demand location. SNG is combined with carbon capture and recycled to minimize carbon emissions to the environment. The SNG synthesis consists of several Sabatier reactors in series run at very high temperatures that range from 250 deg C to 700 deg C; the product is cooled to remove water before SNG is sent to liquefaction.

This paper explores ways to increase thermal efficiency in the SNG Synthesis system. The heat recovery by high-pressure steam is modeled using CHEMCAD simulation software.

154c Capture and Utilization of CO2 in Steel Industry

Abhineet Raj, Naveen Kumar

Current Scenario:

CO_2 is required for GCP water treatment in steel making shops for reduction of TDS(total dissolved solid) from GCP(Gas cleaning plant) water. In this process metallic oxide is converted in to carbonate with help of CO_2 and precipitated out.

Limitation:

1. Presently CO_2 is purchased from suppliers which involve appreciable amount of cost and third-party dependency.
2. In house CO_2 utilization has a potential for reduction of GHG emission

Methodology:

Our company produces slag and part of it is converted to products used in building roads, fly ash bricks etc. and the balance is disposed as waste. Our company intends to utilize this slag waste by carbonization of slag to reduce the disposal costs and to produce value products. For slag carbonation, our company intends to use the CO2 in the stack emissions or any other gas source available within the plant. After diligent study of the gas network with the customer, it is determined to extract the feed gas for the CO2 recovery from the blast furnace and Steel making shop gas mixture network.

Fuel gas for CO2 recovery plant is tapped from the gas network. In this process, all impurities like SOX, NOX, and particulate matters are removed from blast furnace gas in pre-treatment section and after that gas is fed to absorber. In pre-treatment section, sodium bicarbonate is used as scrubbing solution. CO2 is adsorbed in amine (CDRMax solvent) in atmospheric and temperature condition from BF gas and CO2 free BF gas is again sent back to system. The solvent selectively absorbs CO2 from the gas mixture. Then CO2 is separated from rich amine solution in stripper at 80-1000C using steam in reboiler and solvent is regenerated as lean solvent. The generated CO2 is 96% pure on wet basis and more than 99% pure on dry basis.

Benefits:

1. CO_2 is separated out from Blast furnace gas and that CO_2 is utilized in GCP water treatment process
2. After CO_2 recovery from Blast furnace gas, rest part of the gas is again sent back to the system
3. Favourable to environment due to reduction of GHG emission, mitigation of climate change through CO_2 capture
4. 1st of its kind to be used in any steel company worldwide
5. Saving of annual costs of 20 million INR.
6. It produces 5 tonnes per day (TPD) industrial plant for carbon capture and utilisation from blast furnace off-gas which helped in reducing freshwater consumption, developing sustainable supply chains, and imbibing circular economy.

Development of a flat-sheet composite membrane for alkaline water electrolysis: a statistical study using response surface methodology

Ruchika Gupta, Arindam Sarkar, Bharat K. Suthar, Akkihebbal K. Suresh
Indian Institute of Technology Bombay, Mumbai, Mh., India

Introduction

Asymmetric polymeric membranes are synthesized using phase inversion technique. In this process, the interaction of polysulfone with N-Methyl-pyrrolidone (NMP) and water affects the thermodynamics and the kinetics of the membrane's morphology and performance, which has been explored in previous literature[1,2]. Polysulfone, NMP and water interaction is described by a ternary phase diagram that explains the process of phase inversion where coagulation of polymer takes place, and depending upon the conditions, structures are formed. By adding a non-solvent additive to the dope solution, a defect-free (less macro-voids and surface holes) membrane can be formed. It can decrease the dissolving power of solvent and increase the polymer-polymer interaction, resulting in rapid solidification at membrane surface, forming a thinner skin layer and more uniform structure. The zirfon membranes include additional zirconia particles to the PSf-NMP-Water system which brings in an additional factor that can influence the structure. Thus, a detailed analysis of the synthesis procedure has been studied here through the design of experiments. The porous structure of zirfon is greatly affected by two synthesis characteristics: the ratio of zirconia to polysulfone and the rate of liquid-liquid demixing. The membrane structure becomes denser with increasing zirconia loading, reducing the volume of finger-like pores when compared to a porous-polysulfone membrane[3,4]. In this context, based on the preliminary results of bubble point pressure measurements, the performance of Zirfon membranes may be improved through structural modifications including the dense layer by optimizing the polymer concentration and the rate of solvent extraction in the membrane preparation process.

In the process of establishing steps of a fabrication procedure for membranes, since various parameters affect the morphology of the membranes and its performance, it is difficult to assess how much the response variable changes with each factor and how the factors interact to produce different results. Thus, a statistical design of experiments in conjunction with response surface methodology helps in planning the experiments, analyzing the effect of a number of parameters simultaneously by statistical methods.

Experimental Methodology

Materials

The materials used for the fabrication of membrane samples for this study were Polysulfone polymer beads of average molecular weight 22,000 purchased from Sigma, zirconium dioxide (calcined), Mw = 123.22 in powder form purchased from Sigma Aldrich, N-Methyl-2-pyrrolidone (NMP) from Sigma Aldrich, Tetrahydrofuran (THF) from Sigma Aldrich, zirconium dioxide (calcined), Mw = 123.22 in powder form purchased from Sigma Aldrich, India. Deionized water was used throughout the membrane preparation process.

Methodology

Membrane preparation

The membranes were prepared by using the solution casting and phase inversion method, shown in figure 1. For the preparation of membranes, zirconium dioxide, and polysulfone were taken in the weight ratio of 80:20, with their total: solvent as 40:60[5,6],. A suspension of finely ground zirconia (zirconium dioxide) powder in NMP was prepared by sonication, and then polysulfone beads were added slowly to the solution over a duration of 8-10 hours, the suspension being kept under rigorous stirring at 400 rpm at room temperature, until all the polymer dissolves and a uniform dope suspension is obtained. The stirring speed is reduced to 200 rpm and stirring continued for the next 12 hours for the final dope, ready for membrane casting.

The steps involved in the synthesis of a porous flat-sheet zirfon membrane are as follows:

1. Preparation of dope solution (explained above) with polymer and zirconia in solvent
2. Casting of the solution (with or without reinforcement support) as a flat sheet using a film applicator
3. Solvent evaporation by air drying for dense layer formation
4. Solvent extraction by an anti-solvent by wet phase inversion for the formation of the porous sub-structure
5. Hot pressing of the membrane in case of membranes with mesh fabric support for better integration.

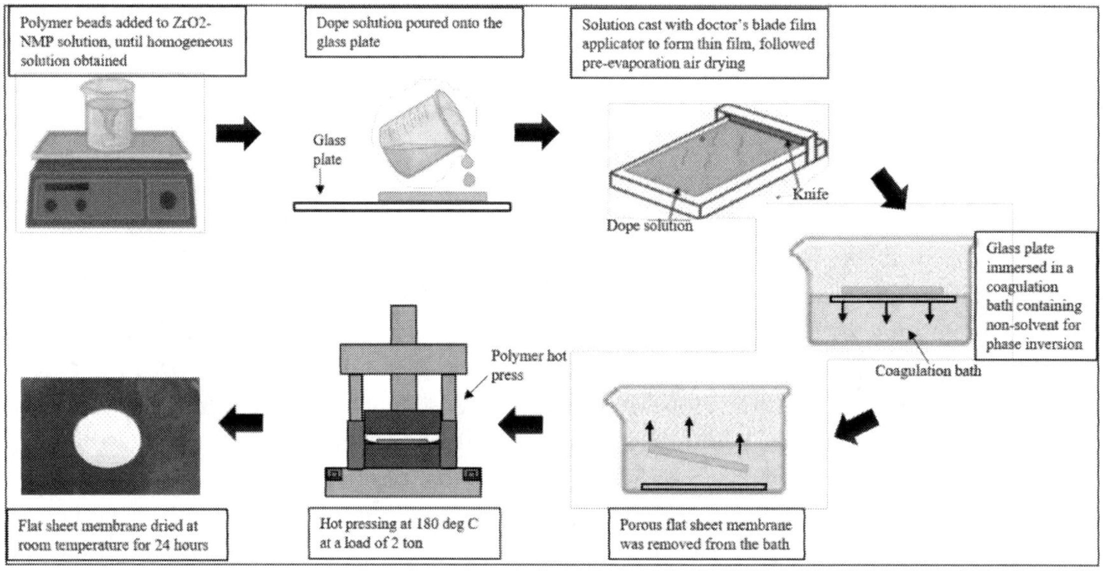

Figure 1. Membrane preparation by solution casting and phase inversion method

Experimental Design

A central composite experimental design (CCD) was used, as generated by the statistical software Design Expert version 11. The selection of factors for the experimental design is done by prioritizing the properties required in a membrane separator suitable for alkaline water electrolysis. Membranes that need to be used in flow electrolyzers with liquid electrolytes should possess characteristics of minimum resistivity (ρ) for low ohmic drop and minimum gas crossover characterized by a high bubble point pressure.

1. Bubble point pressure

The bubble point pressure of the membranes has been tested using a gas-permeation setup. The measurements were carried out by pressurizing one side (feed side) with air in a gas-permeation cell, in which a membrane soaked with water is installed. The flow of air can be controlled by a pressure regulator. The permeate side of the cell is connected to a bubble point apparatus and the bulb is filled with soap solution. As the pressure of air is increased, at some point bubbles rise in the bubble column. Appearance of the first bubble is taken as the bubble point pressure in the pressure gauge.

2. Resistivity measurements

The resistivity of membranes was measured using the CR2032 coin cells made up of SS316 by the Electrochemical Impedance Spectroscopy method using the PARSTAT 3000A potentiostat, in 0.5M NaOH solution. Impedance data were recorded in a frequency range of 100kHz to 1Hz and an AC rms voltage of 10 mV, and the high-frequency resistances were used to calculate the membrane resistivity.

Experiments

Screening experiments

A screening of the various factors affecting the structure and performance of the membrane is done to identify the design variables that have a significant effect for further investigation. The effect of the addition of THF as a co-solvent with NMP solvent in the dope solution on the morphology of membranes was investigated, prior to the study. A significant reduction in the size of enlarged finger-like pores was observed. A reduction in macro-voids was also found, as shown in the SEM images. The surface of the membranes also showed a reduction in the size and number of pores (holes). Screening experiments were carried out with these factors to confirm the significant impact on the responses. Figure 2 shows the resistivity, sorption, and bubble point pressure of the membrane produced at different NMP to THF ratios and oven drying times.

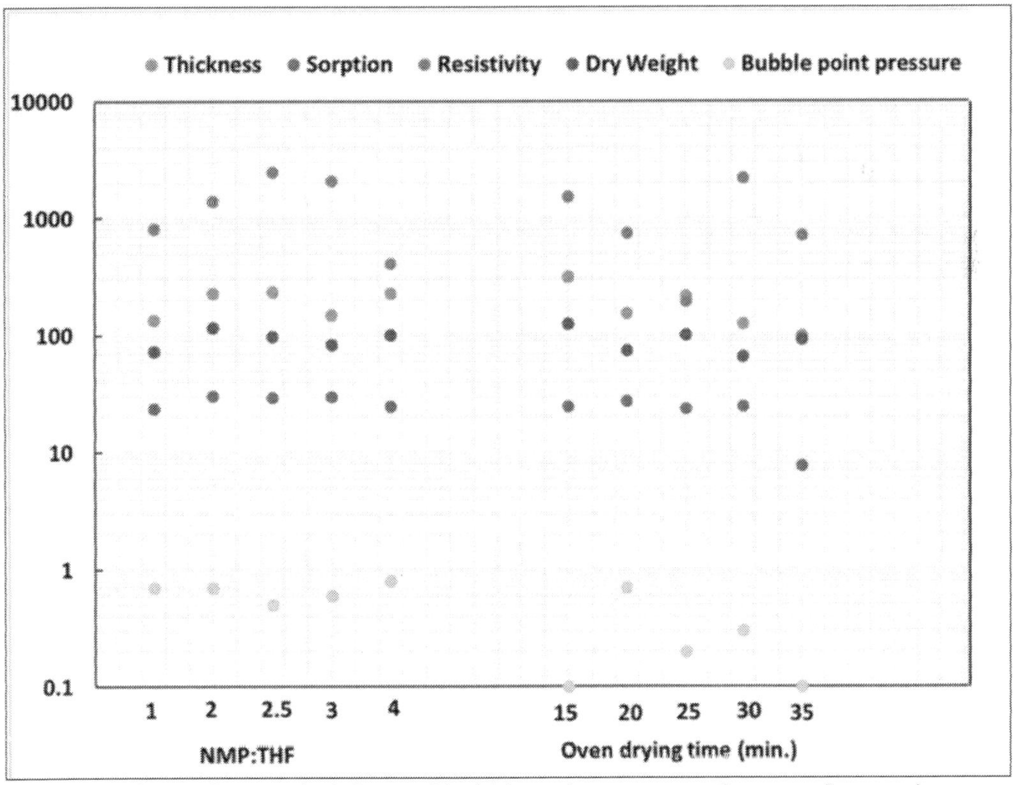

Figure 2. Output of sorption, resistivity and bubble point response factors of screening experiments with THF concentration and oven drying time as input factors

Results and discussion

Response surfaces

The experiments were carried out as a result of the CCD method; resistivity and bubble point responses were measured. The responses were analyzed through the models predicted, ANOVA and response surfaces methodology. The response surfaces and contour plots for resistivity and bubble point pressure are shown in Figures 3 and 4. In Figure 3, three response surfaces for resistivity have been shown with factors A and B, at C (PSf conc.) 8%, 10%, and 12%. For minimum membrane resistivity, the blue zones can be considered, where the resistivity increases with THF conc. and decreasing with oven drying time. The surface plots for bubble point pressure have been shown at oven drying times as 15min, 25min, and 35 min. For a maximum bubble point pressure of the membrane, we consider the maximum point on the curved surface, color depicted as light green to yellow in the response surfaces and the contours. In the surface plots for bubble point, shown in Figure 4, the PSf concentration decreases with the reducing OD time for a high bubble point region for the same range of THF concentration. For a high bubble point pressure (4 bar approximately), a high OD time of 35 minutes and a PSf concentration from 8% to 9% with a THF range from 2.5% to 4% is predicted.

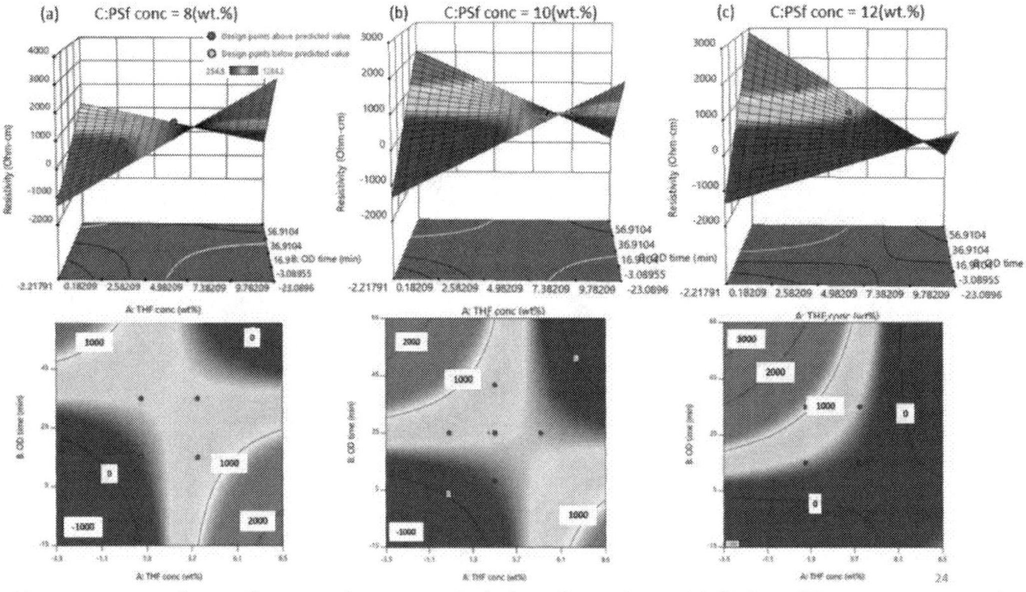

Figure 3. Response surfaces for membrane resistivity plotted on (a) Polysulfone concentration = 8%, (b) Polysulfone conc.=10%, (c) Polysulfone conc.= 12% with their respective contour plots

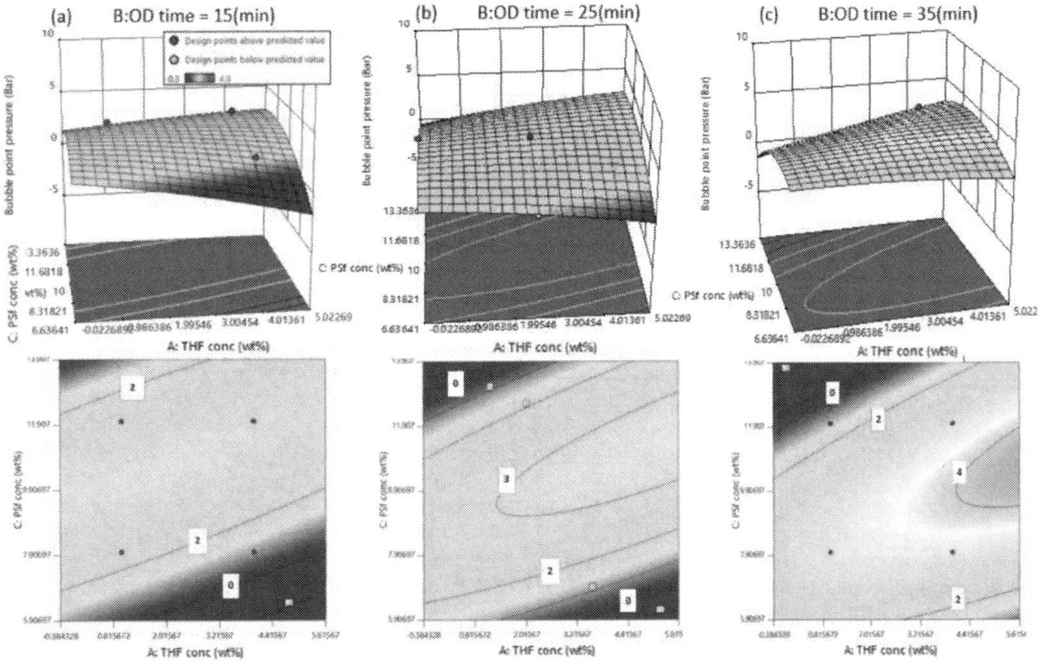

Figure 4. Response surfaces for membrane bubble point pressure plotted on (a) oven drying time = 15 min, (b) oven drying time = 25 min (c) oven drying time = 35 min with their respective contour plots

Conclusions

In this work, varying the polymer (polysulfone) concentration, oven drying time in solvent evaporation and using THF as a co-solvent for NMP solvent in membrane fabrication procedure have been studied to observe the structural changes in the membrane and its performance in resistivity and gas crossover, for a physical separator suitable for application in alkaline electrolysis. The addition of tetrahydrofuran (THF) as a co-solvent to NMP solvent shows a significant morphological change as expected. The membranes were evaluated by measurements of bubble point pressure and resistivity. Membranes prepared with THF and higher polymer concentration exhibited higher bubble point pressure. A response surface methodology has been used to optimize the concentration of THF, time of oven drying and polysulfone concentration in the steps involved in membrane synthesis. The mathematical models proposed and verified using diagnostic plots, describe the responses within the ranges of factor investigation. An optimal window in the parameter space based on the response region of minimum resistivity and maximum bubble point, can be proposed as: THF conc. between 2.5 to 5, oven drying time between 25 to 41.8 and Polysulfone conc. between 10% to 12%.

References

(1) Aroon, M. A.; Ismail, A. F.; Montazer-Rahmati, M. M.; Matsuura, T. Morphology and Permeation Properties of Polysulfone Membranes for Gas Separation: Effects of Non-Solvent Additives and Co-Solvent. *Sep. Purif. Technol.* **2010**, *72* (2), 194–202.

(2) Mohsenpour, S.; Safekordi, A.; Tavakolmoghadam, M.; Rekabdar, F.; Hemmati, M. Comparison of the Membrane Morphology Based on the Phase Diagram Using PVP as an Organic Additive and TiO2 as an Inorganic Additive. *Polymer (Guildf).* **2016**, *97*, 559–568.

(3) Vermeiren, P.; Moreels, J. P.; Leysen, R. Porosity in Composite Zirfon® Membranes. *J. Porous*

Mater. **1996**, *3* (1), 33–40.
(4) Vermeiren, Philippe; DOYEN, Wily; ADRIANSENS, Walter; LEYSEN, R. Preparation of a Membrane and Its Use in an Electrochemical Cell, 1993.
(5) Xu, L.; Li, W.; You, Y.; Zhang, S.; Zhao, Y. Polysulfone and Zirconia Composite Separators for Alkaline Water Electrolysis. *Front. Chem. Sci. Eng.* **2013**, *7* (2), 154–161.
(6) Vermeiren, Ph.;Adriansens, W; Moreels, J.P.; Leysen, R. Evaluation of the ZirfonS Separator for Use in Alkaline Water Electrolysis and Ni-H2 Batteries. *Int. J. Hydrogen Energy* **1998**, *23* (5), 321–324.

60b Engineering the Cyanobacterial Photosynthetic Electron Transport Chain to Improve Photosynthetic Efficiency.

Sumant Brahmankar, Nima Hajinajaf, Aditya Sarnaik, Jackson Comes, Parker Poole, Christine Lewis, Christine M Lewis, Nandini Kannoju, Danika Kartchner, Jonathan Nguyen, Willem Vermaas, Abhishek Singharoy, Arul Mozhy Varman

Solar energy harvested by cyanobacteria is a primary source for energy. Therefore, it is imperative to increase its photosynthetic efficiency to secure our food supply. To this end, the cyanobacterium *Synechocystis* sp. PCC 6803 (hereafter, *Synechocystis*) serves as a perfect model organism as it is easy to engineer and shares all energy bottlenecks ubiquitous to photosynthesis and respiration in plant cells. In this study, we designed two distinct strategies to engineer the photosynthetic electron transport chain in *Synechocystis* to increase the photosynthetic efficiency. The first approach was directed towards linking the photosynthetic complexes photosystem I (PSI) and cytochrome *bf* (cyt*b6f*), to bring them in a closer proximity and thereby reduce electron loss. We performed computational 3D protein analysis to identify suitable proteins to link cyt *b6f* and PSI. Comprehending the protein orientations and interpeptide distances PetA from cyt *b6f* (C-terminus) and PsaA from PSI (N-terminus) were identified as suitable candidates to be linked. Following these strains expressing the linked proteins were constructed for further experiments. In the second approach to increase the membrane density of cyt *b6f* complex and thereby reduce the proximity between cyt *b6f* with PSI and PSII, strains were constructed to overexpress the major proteins PetA, PetB, PetC, and PetD. The hypothesis was that overexpressing one of the major proteins would result in increasing the density of cyt *b6f* by leveraging unknown endogenous regulatory systems. Comparative growth studies of the engineered strains with wild type under atmospheric CO2 showed that at low light (20 μE m−2 s−1), there was a 52% and 39% increase in cell density of the strain expressing PetD and Linker compared to the WT at day 12, respectively. Interestingly, this difference in growth at moderate light (80 μE m−2 s−1) was 34% and 15% for strain expressing PetD and Linker compared to WT at day 11. Experiments under atmospheric CO2 and high light (150 μE m−2 s−1) also demonstrated an extended growth pattern with 63% increase in biomass production for PetD and Linker strains compared to WT by day 20. Joliot type spectroscopy measures the activity of PSI's ability to complete electron flow in the photosynthetic electron transport chain also highlights the PetD with a higher electron flow than both WT and the Linker at all light intensities (20, 45, 80, 150, 320 and 2050 μE m-2 s-1). The Linker strain further showed more rapid electron flow than WT only at light intensities of 20, 45, and 80 μE m-2 s-1. These findings confirm our hypothesis that in the engineered PetD and Linker strains, the electrons were channeled more effectively through the photosynthetic electron transport chain and thereby, improved photosynthetic efficiency which can be translated into higher growth rate.

60c Conversion of Coal Tar into Needle Coke for Utilization in Graphite Electrode and Li-Ion Battery

Naveen Kumar, Abhineet Raj

BACKGROUND:

This innovation is on conversion of coal tar into needle coke which is the main raw material for graphite electrode & Li-ion battery. Presently, 1.8 lakh tons of needle coke is imported in India worth Rs. 3000-3500 Cr. annually. Our company produces 2 lakh tons of coal tar which will be increased to 3 lakh ton by 2030. Coal tar can be converted into many high-end carbon products such as QI free pitch, binder pitch, carbon foam, carbon fiber, meso carbon micro beads (MCMB), anode materials for Li-ion batteries, and needle coke. Needle coke was selected because it has got three major applications: graphite electrode, synthetic graphite (anode materials for LIB), carbon raiser in LD. After successful lab work, we have put up a pilot plant of 1 TPA capacity. Initial testing of the samples shows that properties (carbon content, density, sulphur, volatile matter, ash content) of our needle coke are close to the imported needle coke properties. This project has a potential of ₹ 3000 Cr. new business for our company.

UNIQUENESS:

Coal tar is converted into pitch through distillation and then to needle coke through carbonization process. Needle coke technology is closely guarded, and the process has many know-how/trade secrets to get the desired quality of coke, its needle like structure and other properties. The yield and structure of the final product is very important in deciding the quality of the final product and economics. Typical coal tar distillation gives 50% oil and 50% pitch. Around 25% of anthracene oil (AO) can be distilled off from the distillate oil. This AO, if used directly along with pitch then it will vaporize at high temperature during coking operation. To solve this issue, we polymerized the AO and pitch mixture in presence of catalyst before feeding it to coker unit. The catalyst accelerated dehydrogenation during the polymerization of coal tar pitch and AO. This polymerized pitch is carbonized to get the needle coke. This process increases overall yield up to 5% (from 20 to 25%). Second innovation is on coker design. We designed the coker unit in such a way that it solved many designs related issues.

BUSINESS CONNECT:

Entire amount of needle coke is imported in India. Because of supply chain issue, HEG and Graphite India keep huge amount of inventory. Coal tar produced in BPP can be converted to value added products through this technology. Ultimately this technology will benefit the India in reducing the CO_2 emissions as well as making India green energy hub

by following way:

(1) needle coke is main raw material for graphite electrode manufacturing. Graphite electrodes are used in electric arc furnaces (EAF) for steel making. In coming decades, more steelmaking capacity will be added through EAF route to reduce the CO_2 emissions.

(2) The needle coke can be converted to anode materials for Li-ion batteries (LIB). So, this technology will make the prime raw materials (for EAF and LIB) available in India.

IMPACT:

New business of ₹ 3000-3500 Cr. for our company

This technology has potential to produce different carbon materials

It will accelerate green energy adoption in India (waste to wealth)

137a Nano and Single-Atom Catalysts for Biomass Conversion to Renewable Liquid Fuels

Maoqi Feng

Catalysts play a vital role in chemical and energy industries. Shifting fossil energy towards renewable fuels has created great demand for catalysts development. Nano-catalysts and single-atom catalysts (SACs) have unique catalytic properties for the rational design of new catalysts with high activity, selectivity, stability and most importantly better atom economy.

Nano-catalysts and SACs manufacturing is very challenging, especially for metals with agglomeration at higher temperature such as copper and ruthenium. Recently, we developed an inexpensive low-temperature plasma (LTP) device for the preparation of nano-catalysts and SACs, including bimetallic nano-catalysts. The bimetallic nano-catalyst Ru-Cu/zeolite and SACs Ru/Cu-zeolite catalyst were tested for lignin conversion. Hydrocarbon fuels were obtained with high yield of biomass conversion (over 85%).

The LTP technology greatly improves catalysts manufacturing efficiency and performance through innovations by 1) reducing catalyst calcinations and activation temperature and time, 2) minimizing metal agglomeration leading to more uniform distribution of bi/multi- metal particles, and 3) enhancing anchoring of metals on the support for more robust and longer catalyst life.

In summary, we will present a technology we are developing, which is promising and environmentally friendly LTP tool for the manufacturing of bimetallic nanocatalysts and SACs and can be used for biomass conversion.

Acknowledgement

This project was supported by DOE SBIR Project Contract Number DE-SC0022796.

137b Quantification of Biomass Pyrolysis Products Using GC-MS/PA-FID Analysis

Cara Kilmartin, Melinda Barron

This study seeks to identify and quantify biomass pyrolysis products in tangent through the use of dual gas chromatography (GC) detectors. The GC uses mass spectrometry (MS) to identify components in the biomass product and a flame ionization detector (FID) coupled with a Polyarc (PA) methanation reactor to readily quantify carbon present in each sample. The PA reactor converts every carbon atom present in each compound to methane prior to reaching the FID; therefore, the FID response factors for each compound are the same and only one calibration standard is needed to quantify the entire range of bio-oil products. There are limitations on what types of products can be detected by the GC, including non-volatiles, high molecular weights, highly polar compounds, etc. The GC-MS/PA-FID sheds light on what percentage of products are not analyzed by the GC and allows for reporting of absolute rather than relative product amounts.

A 2.5" fluidized bed reactor was used to pyrolyze a loblolly pine feedstock at temperatures from 450 – 500 C. Both organic and aqueous liquid biomass pyrolysis products were collected and analyzed using the GC-MS/PA-FID system. Each detectable compound present in the sample was identified, and the mass of carbon in each detectable compound was determined. The sum of carbon in all compounds gives the total detected carbon in the bio-oil product. The total carbon present in each collected sample was measured from CHN elemental analysis; therefore, the percent of carbon analyzed can be found. Carbon balances of different bio-oil samples were analyzed at various reaction conditions and correlated to the water content in the samples. Trends between sample composition and carbon analyzed were observed and have led to greater insight into biomass pyrolysis products.

Biodiesel Synthesis From Oil Having High Free Fatty Acid Content Using Low-Cost Heterogeneous Catalysts

Vishnulal Kakkaraveetil Sukumaran[a], Venkatasailanathan Ramadesigan[a], Srinivas Seethamraju[a,1]

[a] Department of Energy Science and Engineering, Indian Institute of Technology Bombay, Powai, Mumbai, 400076, India

Environmental pollution due to carbon emissions is increasing worldwide due to increased crude oil consumption to meet energy demands. Energy production from renewable energy sources and its proper use is essential to reduce environmental pollution, increase energy security and lower fossil fuel consumption. Some renewable energy resources are solar, wind, geothermal, and biofuels. Biodiesel is one type of biofuel that can be produced locally. Biodiesel is a mixture of different fatty acid alkyl esters (FAAE), which can be used as a fuel-blend in diesel engine vehicles and reduces CO_2 emissions [1]. The chemical reactions involved in biodiesel synthesis include the transesterification of triglycerides (TGs) and esterification of free fatty acids (FFA) with alcohol (methanol or ethanol). Both are homogeneous or heterogeneous catalytic reactions; a base catalyst for transesterification and an acid catalyst for esterification are required. In comparison, the kinetics of transesterification is faster than the esterification at low reaction temperatures (T = 65 °C) when a base catalyst is used. So, transesterification of TGs is preferred over esterification of FFA to prepare the biodiesel commercially. Nowadays, edible oil containing 100 % TGs is used for commercial biodiesel synthesis – however, this leads to cooking oil scarcity and a food-versus-fuel debate [2].

The production cost of biodiesel is higher than petroleum diesel (about 3 times higher) if edible vegetable oil is used as a feedstock. The production cost can be significantly reduced using low-cost feedstocks like non-edible vegetable oil and waste cooking oil. It would solve the environmental problems arising from waste cooking oil disposal [3]. However, these feedstocks have higher FFA content (>2 wt.%), resulting in soap formation with the base catalyst during the transesterification of TGs in the oil. So, the TG to biodiesel conversion is adversely affected due to the consumption of the base catalyst. Also, the presence of soap reduces the biodiesel quality [4]. The alternative way of avoiding the saponification of FFAs is to reduce the FFA amount in the oil by performing the esterification first. After that,

[1]Corresponding Author: Department of Energy Science and Engineering, Indian Institute of Technology Bombay, Powai, Mumbai 400076, India; s.srinivas@iitb.ac.in

transesterification is performed to convert the remaining TGs to biodiesel. This two-step method of biodiesel synthesis increases the overall production costs. If a homogeneous catalyst (base or acid) is used, then water washing of biodiesel is recommended after the synthesis. Water washing of biodiesel is required to remove homogeneous catalysts and soap from the biodiesel, which causes a lot of wastewater generation [5].

To overcome these problems, synthesizing a heterogeneous solid catalyst containing both acidic and basic sites (bifunctional catalyst) is recommended, which can be separated after reaction easily through simple filtration. The bifunctional catalyst can help both the transesterification and esterification of TGs and FFAs simultaneously in a single pot at the same reaction conditions. The bifunctional catalysts can be synthesized by impregnation, sol-gel and co-precipitation methods [6]. It is noted that there are only a few works reported on using a physical mixture of acid and base catalysts for the synthesis of biodiesel from oil containing higher FFA.

A physical mixture of two low-cost heterogeneous catalysts is used to synthesize biodiesel from high FFA (5 wt.%, 10 wt.% and 20 wt.%) containing oil in this work. The catalysts used are an acid catalyst in the form of a silica-alumina-based compound (SiAl) from an industry and base catalysts CaO and calcium diglyceroxide (CaD). The CaD catalyst is synthesized by refluxing the CaO, glycerol and methanol in a mass ratio of 1:14.667:3.33 at 60 °C [7]. The CaO is synthesized by calcining the waste eggshell powder (particle size <500 μm) at 900 °C for 3 hours. The catalyst SiAl is also calcined at 900 °C to remove the unwanted impurities from the surface. The catalysts used are derived from waste materials and help in the efforts to move towards a circular economy for biodiesel synthesis. The proportions of CaO and SiAl mixing used were CaO + 5 % SiAl, CaO + 10 % SiAl, and CaO + 20 % SiAl. The biodiesel synthesis was carried out at reaction conditions of 65 °C, 3 wt.% of catalyst loading, 6:1 methanol-to-oil molar ratio and 200 rpm for the stirring. The reaction setup includes a round bottom flask (RBF) connected with a condenser; mixing and heating the reactants in the RBF were carried out by a mechanical stirrer and an oil bath, respectively. The structural and physio-chemical characterizations of the catalysts were studied using X-ray diffraction (XRD), scanning electron microscopy (SEM), and Fourier transform-infrared spectroscopy (FT-IR) techniques. The biodiesel conversion and ester content present in the biodiesel are analysed using proton nuclear magnetic resonance spectroscopy (^{1}H-NMR) and gas chromatography coupled with a flame ionization detector (GC-FID), respectively. A thermogravimetric method was developed to quantify the soap formation from FFA. The solid residue (SR) of the reaction was collected through filtration. The SR contains the catalyst and soap. The amount of soap in the SR was quantified by the thermogravimetric method. It helped us to understand the performance of the SiAl acid catalyst. The calcium amount in the biodiesel due to the leaching of Ca^{2+} ions was determined by the inductively coupled plasma atomic emission spectroscopy (ICP-AES).

The thermogravimetric analysis of SR collected after biodiesel synthesis from oil containing 5 % FFA with different catalyst systems (CaO + x % SiAl, x = 5, 10, and 20) shows that the presence of SiAl catalyst significantly reduced saponification of FFA with CaO. It increased the overall conversion of oil into biodiesel. The lowest FFA to soap conversion was obtained for the catalyst mixture of composition CaO + 20 % SiAl; only ~20 wt.% of FFA was converted into soap, and the remaining FFA was converted into biodiesel via esterification. The FFA to soap conversion was ~60 wt.% and ~30 wt.% for CaO + 5 % SiAl and CaO + 10 % SiAl, respectively. The reduction in saponification of FFA increased the availability of CaO to catalyse the transesterification faster. So, the total reaction time taken to get maximum conversion was reduced compared with a transesterification of the same feedstock by CaO catalyst. The total reaction time taken for obtaining >80 % conversion was 14 hours for the reaction catalysed by CaO without SiAl catalyst. However, the reaction time was reduced to 7 hours when a mixture of CaO and SiAl catalyst was used, and the conversion reached above 99 %. The performance of the CaD + 20 % SiAl catalyst system was studied in the biodiesel synthesis from oil containing 5 wt.%, 10 wt.% and 20 wt.% FFA. The substitution of CaO by CaD in the catalyst mixture further reduces the reaction time to 1 hour with minimum soap formation. The catalyst loading was 3 wt.%, 6 wt.% and 12 wt.%, respectively for 5 wt.% FFA, 10 wt.% FFA and 20 wt.% FFA contained oil. The FFA to soap conversion was ~20 wt.% for 20 wt.% FFA case and ~10 wt.% for both 5 wt.% FFA and 10 wt.% FFA cases. The GC-FID analysis of the synthesized biodiesel confirmed the presence of methyl esters of palmitic acid, oleic acid, and linolenic acid in the biodiesel.

This study concludes that biodiesel was produced from a higher FFA-contained oil (5 wt.%, 10 wt.% and 20 wt.% FFA) using a physical mixture of two low-cost heterogeneous solid catalysts, base catalysts CaO/CaD and an acidic SiAl catalyst. The presence of SiAl helped to achieve a higher conversion of biodiesel from higher FFA-contained oil with less soap formation.

[1] Thangaraj B, Solomon PR, Muniyandi B, Ranganathan S, Lin L. Catalysis in biodiesel production—a review. Clean Energy 2019;3:2–23. https://doi.org/10.1093/ce/zky020.

[2] Khan TMY, Atabani AE, Badruddin IA, Badarudin A, Khayoon MS, Triwahyono S. Recent scenario and technologies to utilize non-edible oils for biodiesel production. Renew Sustain Energy Rev 2014;37:840–51. https://doi.org/10.1016/j.rser.2014.05.064.

[3] Pugazhendhi A, Alagumalai A, Mathimani T, Atabani AE. Optimization, kinetic and thermodynamic studies on sustainable biodiesel production from waste cooking oil: An Indian perspective. Fuel 2020;273:117725. https://doi.org/10.1016/j.fuel.2020.117725.

[4] Verma P, Sharma MP. Review of process parameters for biodiesel production from different feedstocks. Renew Sustain Energy Rev 2016;62:1063–71. https://doi.org/10.1016/j.rser.2016.04.054.

[5] Atadashi IM, Aroua MK, Aziz AA. Biodiesel separation and purification: A review. Renew Energy 2011;36:437–43. https://doi.org/10.1016/j.renene.2010.07.019.

[6] Mansir N, Teo SH, Rashid U, Saiman MI, Tan YP, Alsultan GA, et al. Modified waste egg shell derived bifunctional catalyst for biodiesel production from high FFA waste cooking oil. A review. Renew Sustain Energy Rev 2018;82:3645–55. https://doi.org/10.1016/j.rser.2017.10.098

[7] Hsiao M-C, Chang L-W, Hou S-S. Study of Solid Calcium Diglyceroxide for Biodiesel Production from Waste Cooking Oil Using a High Speed Homogenizer. Energies 2019;12:3205. https://doi.org/10.3390/en12173205

116a Induction Heating for the Ammonia Cracking Process

Cara Kilmartin, Francesco Aimone

The electrification of industrial processes for compatibility with renewable energy has become a major area of interest, both commercially and within the Department of Energy. Process heating comprises more than half of the direct energy consumed by industrial processes and represents a massive opportunity for industrial electrification. Electric induction is a promising method for utilizing renewable energy to heat industrial processes. Due to the increased interest in hydrogen as an energy-dense, low-carbon power and fuel supply, ammonia has received strong consideration as a liquid hydrogen carrier due to its low storage and transportation costs. Ammonia cracking presents a unique opportunity to create on-demand green hydrogen. Hydrogen generated from ammonia cracking can be utilized by commercially available generators, fuel cells, and turbines or consumed in distributed markets including in hydrogen fueling stations, and for chemicals and transportation. For practical and commercial feasibility, these systems must demonstrate the ability to start-up from cold conditions and shut down with minimal latency. The endothermic ammonia cracking reaction can be initiated and sustained by direct induction heating of the catalyst bed; bypassing the heating of the reactor walls reduces the heating time and energy losses, thereby providing an efficient means to intermittently produce hydrogen gas in direct response to demand. The improved efficiency and rapid start-up and shut down from induction heating will enable the use of liquid ammonia as a hydrogen carrier, even at the small-scale.

This study seeks to demonstrate a reduction in energy consumption, faster heating times, and precise temperature control using induction heating. A ruthenium catalyst was used for the ammonia cracking reaction and iron powder used as the susceptor for the induction heating. A series of nine experiments was conducted which test three different temperatures (450 °C, 475 °C, and 500 °C) at three different gas hourly space velocities (GHSVs) (35,000 $h-^{1}$, 50,000 h^{-1}, and 65,000 h^{-1}). The induction experiments were conducted at ambient pressure with a 1:1 susceptor-to-catalyst volume ratio. The time required to reach steady state under each condition, the power input (W), and the hydrogen production were measured to determine the turnover frequency and the efficiency (kWh/mol H_2) for each run. Since iron is an active material for ammonia cracking, a test was run at each experimental condition with the iron susceptor alone to isolate the effect of the catalyst on the calculated parameters. Initial benchmarking of the reaction under the same battery of temperature and GHSV conditions was performed in a conventionally heated microreactor to determine the baseline turnover frequency, start-up times, and power input per mole hydrogen produced for comparison.

116b Optimally Dispersed Network of Biomass Gasification to Produce Green Hydrogen

Yangchuan Xing, Mohsen Dirbaz

In 2023, while hydrogen stands out as a promising carbon-free energy carrier, a staggering 99% of its production remains gray, predominantly derived from natural gas and coal. However, a new paradigm is emerging: green hydrogen production from biomass gasification. This talk delves into the feasibility and advantages of establishing an optimally dispersed network of biomass gasification units across the US. An intriguing economic potential emerges when considering that multiple small-scale biomass gasification units can achieve an economy of scale akin to a single large-scale implementation. Two pivotal arguments underscore the importance of this dispersed network. Firstly, the viability of green hydrogen from biomass escalates as municipalities increasingly recognize efficiencies and capabilities in waste-to-power conversion and other derivatives like chemicals and industrial bulks. The intrinsic ability of hydrogen to stabilize grids, particularly those affected by the intermittent and seasonal outputs of renewables, enhances its value. Secondly, the transportation sector's prospective embrace of hydrogen, with fuel cell vehicles projected to command a 36% market share by 2050, hinges on the success of widespread and accessible green hydrogen production. Establishing an optimally dispersed biomass gasification network can thus play a pivotal role in realizing a sustainable hydrogen-powered future for the US.

116c Intensified Hydrogen Production

Seckin Karagoz

Because of the broad use of fossil fuels for energy generation, the world is experiencing serious environmental issues. As a result, lowering carbon emissions has become a pressing concern. In this context, hydrogen is receiving worldwide attention because of having huge potential such as becoming the leading energy carrier and its mobile/stationary power generating applications, to be part a solution of the aforementioned problem. A significant reduction on the production cost of hydrogen is required to reach the level of widespread use because this is the key factor for the further development and growth of the hydrogen economy.

Developing new technologies or enhancing current processes based on sustainability standards and alternative fuels such as biomass that can substitute for crude oil is crucial to producing clean and cheap power by eliminating the economic and energy penalties. However, the development of sustainable industrial processes is the chemical sector's challenging global goal. The integration of sustainability into process design contributes to the achievement of this target by eliminating or minimizing negative effects.

To develop sustainable chemical, pharmaceutical, and bio-based processes for the future, increased productivity-capacity-safety-flexibility, decreased energy usage-waste-operational costs and simplified process are the main motivation objectives for retrofitting processes. The field of Process Intensification (PI) has a huge potential to satisfy many of the objectives listed above. PI is a rapidly developing field in research and industry that has a huge potential to meet the increasing demands for sustainable production in the chemical process industry. PI has already created many innovations in the chemical process industry with a significant process improvement that is directed toward substantially smaller, cleaner, more energy-efficient technology. Furthermore, PI aims at safer and more sustainable technological developments.

Thus, the main goal of this study is to identify the role of the intensification technologies for hydrogen production processes. The study will consider the following points:

1. Definition of PI and its principles.
2. Listing the existed PI technologies and equipment for intensified hydrogen production.
3. Integration of PI into hydrogen production processes.
4. Implementation of PI technologies for hydrogen production (Current status).
5. Projected status of PI for hydrogen production.

AUTHOR INDEX

AIChE
120 Wall Street, FL 23
New York, NY 10005-4020

ISBN 979-8-3313-1711-9

CURRAN ASSOCIATES INC.
proceedings.com